CONTENTS

Some words are shown in bold, **like this**. You can find out what they mean by looking in the glossary.

HERE'S A PERSONAL STEREO

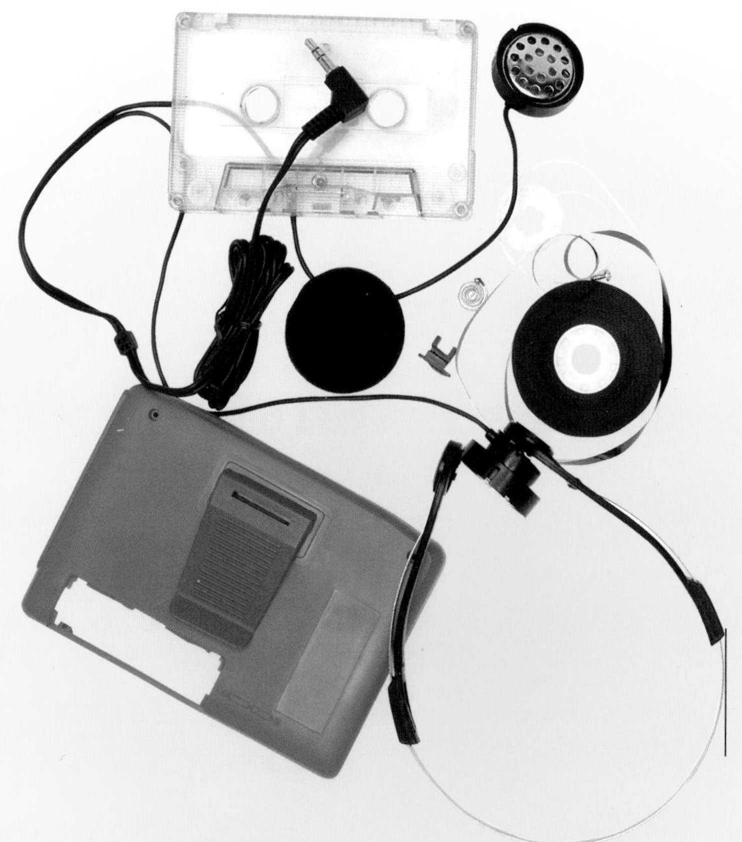

Here are lots of bits
and pieces. There's a coloured
box shape. There's a clear **rectangle**, too.
You can see a long, thin curved piece, a soft
sponge and a hard, shiny battery.

LOOK INSIDE

Personal Stereo

Catherine Chambers

First published in Great Britain by Heinemann Library
Halley Court, Jordan Hill, Oxford OX2 8EJ
a division of Reed Educational and Professional Publishing Ltd
Heinemann is a registered trademark of Reed Educational and Professional
Publishing Limited.

OXFORD FLORENCE PRAGUE MADRID ATHENS
MELBOURNE AUCKLAND KUALA LUMPUR SINGAPORE TOKYO
IBADAN NAIROBI KAMPALA JOHANNESBURG GABORONE
PORTSMOUTH NH CHICAGO MEXICO CITY SAO PAULO

Designed by Celia Floyd
Printed in Hong Kong

03 02 01 00 99
10 9 8 7 6 5 4 3 2 1

ISBN 0 431 08686 9
This book is also available in hardback (ISBN 0 431 08681 8)

British Library Cataloguing in Publication Data

Chambers, Catherine
Personal stereo. – (Look inside)
1. Sound – Equipment and supplies – Juvenile literature
2. Sound – Equipment and supplies – Design and construction
– Juvenile literature
I. Title
621.3'89332

Acknowledgements
The Publisher would like to thank the following for permission to reproduce photographs:
Chris Honeywell, pp.4–21
Cover photograph: Chris Honeywell

Our thanks to Betty Root for her comments in the preparation of this book and to
The Boots Company PLC for their assistance.

Every effort has been made to contact copyright holders of any material reproduced in this
book. Any omissions will be rectified in subsequent printings if notice is given to the Publisher.

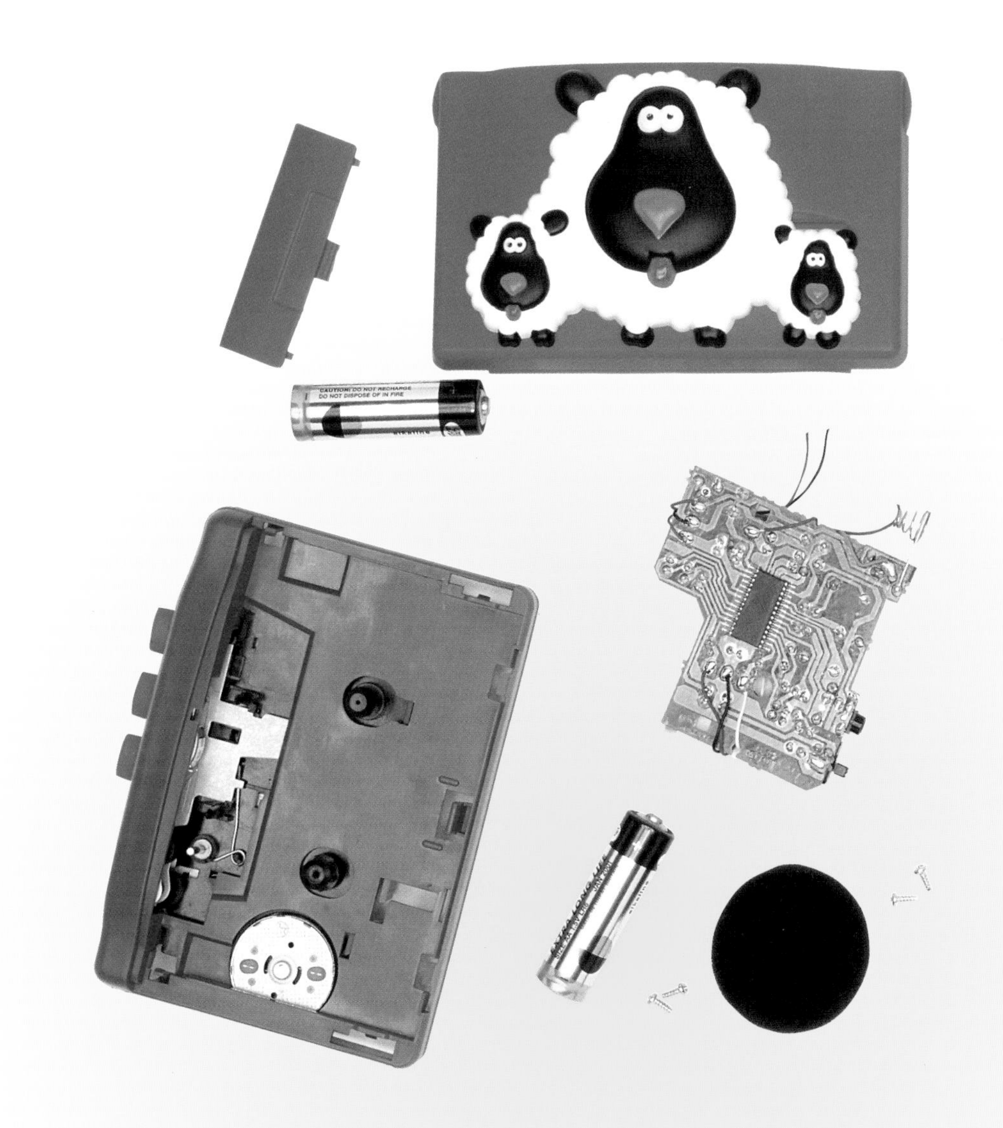

Most of the pieces are tiny. They're all kinds of shapes and are made from different materials. But all the bits and pieces are designed to fit together to make a personal stereo. Here's how!

A HARD CASE

The case is made of tough plastic. The case has to be strong to protect the workings inside it. The workings make the cassette tape turn and play. You can keep the case on your belt with the tough hook.

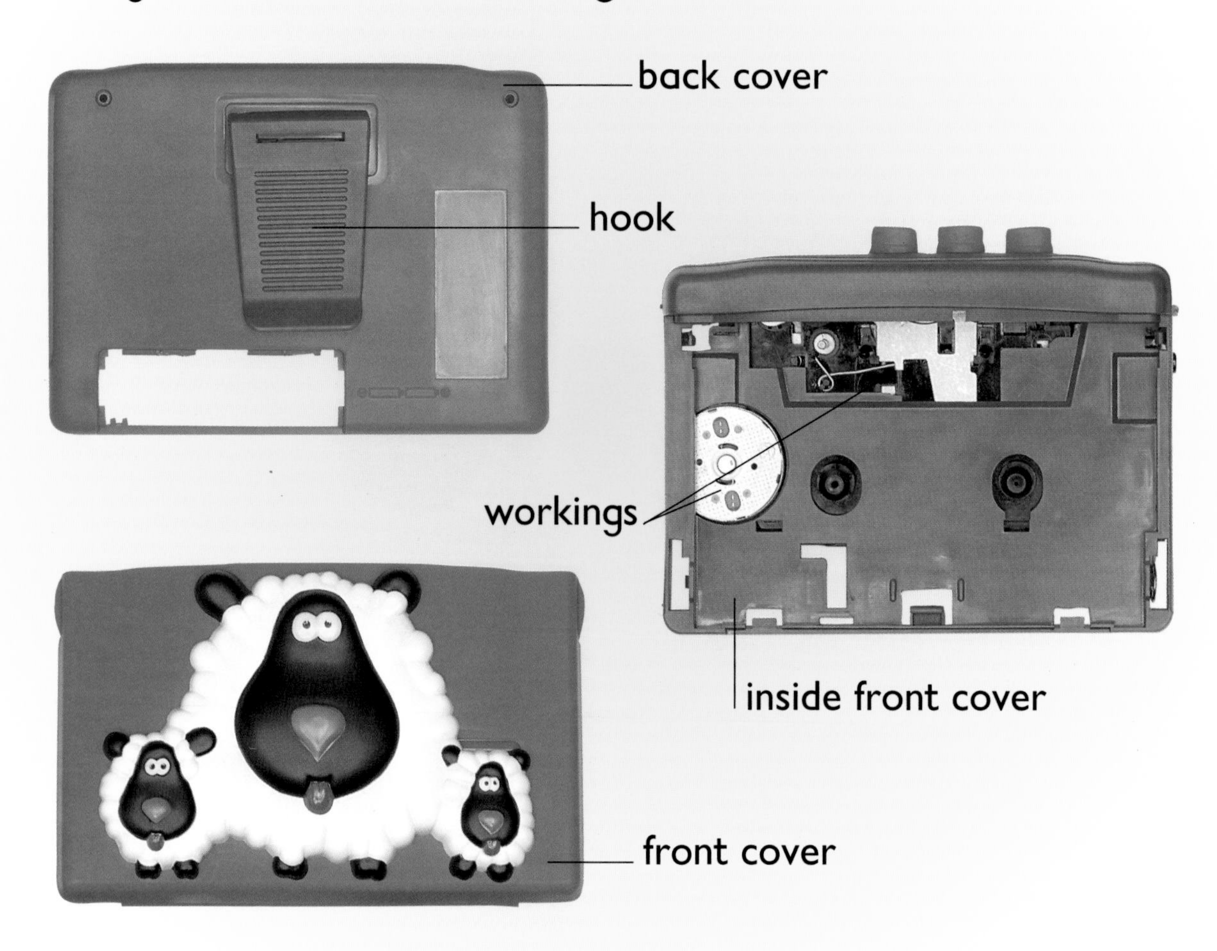

The front cover of the case has **hinges**. These fit into slots in the back cover. The case can open out but the cover won't fall off. There are spaces for the batteries and tape. There's also a cover to stop the batteries falling out.

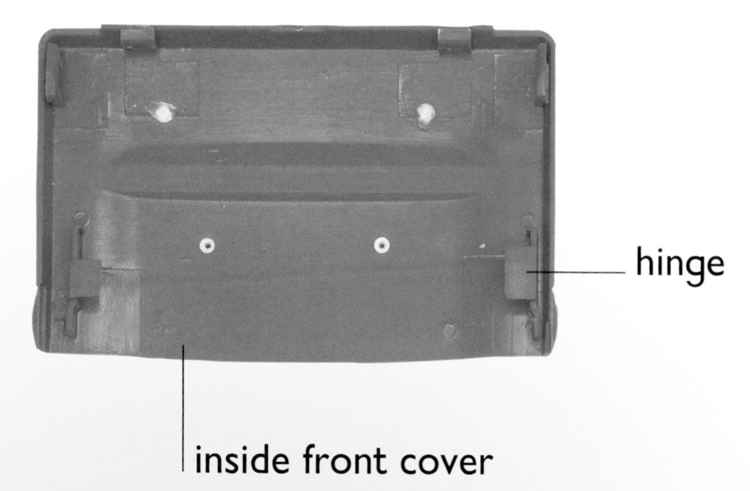

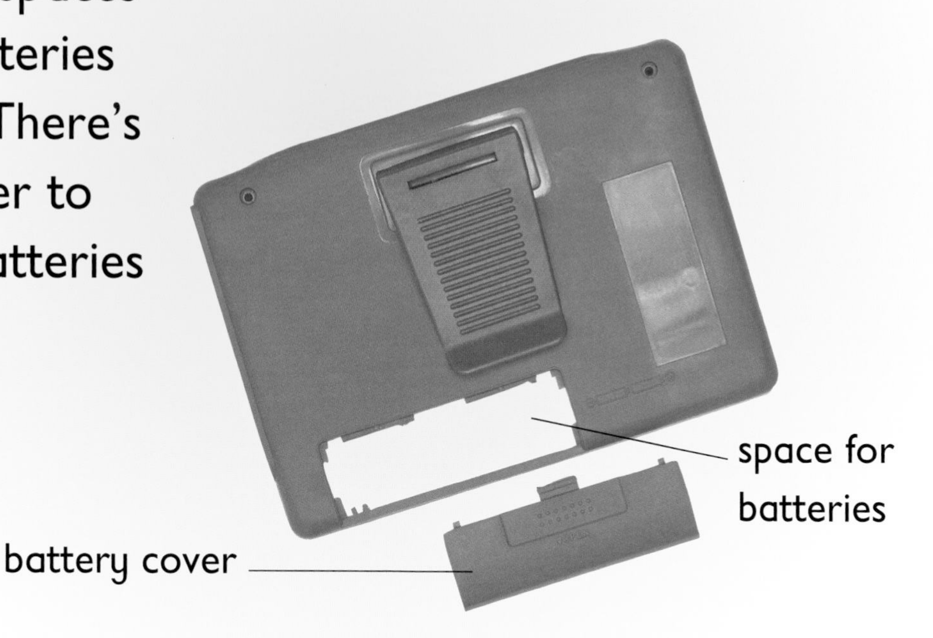

BUTTONS AND SWITCHES

There are three hard plastic buttons on the top of the cassette player. One makes the tape play when it is pushed down. Another moves the tape forward very fast. The third stops the tape.

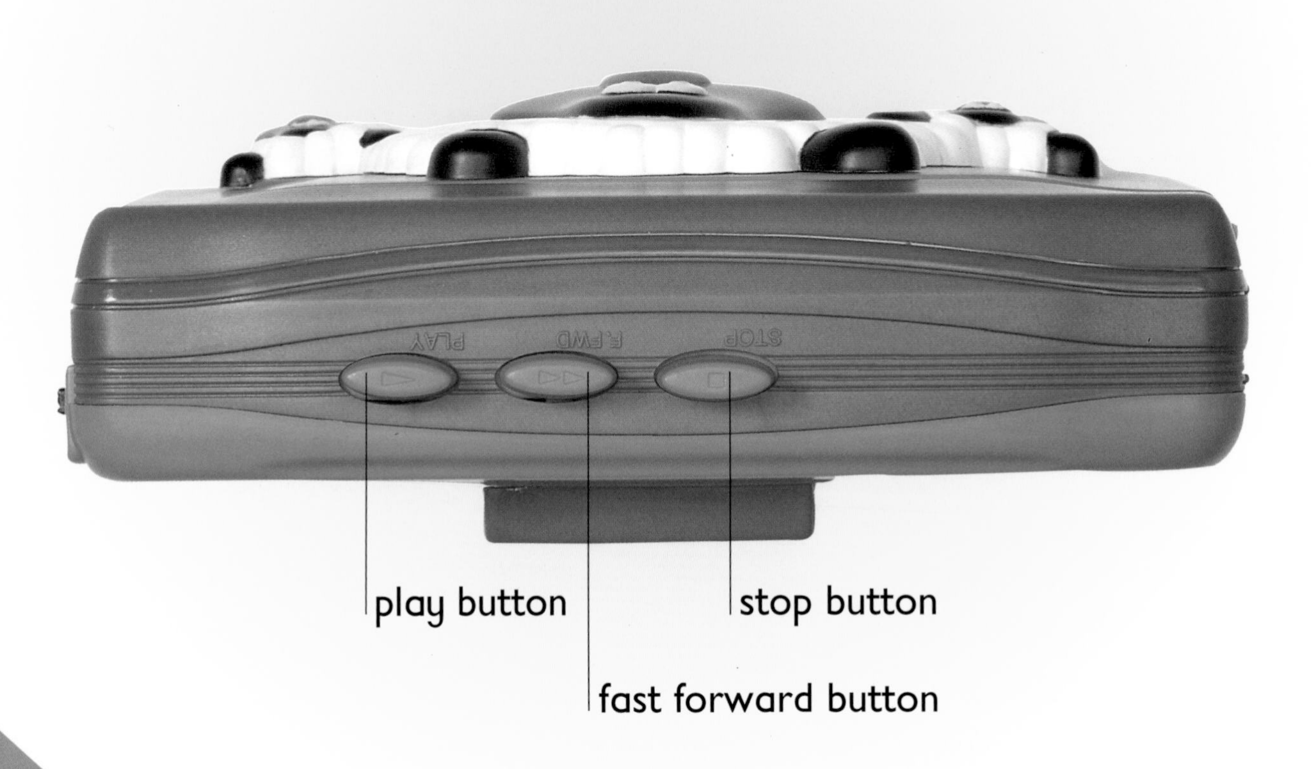

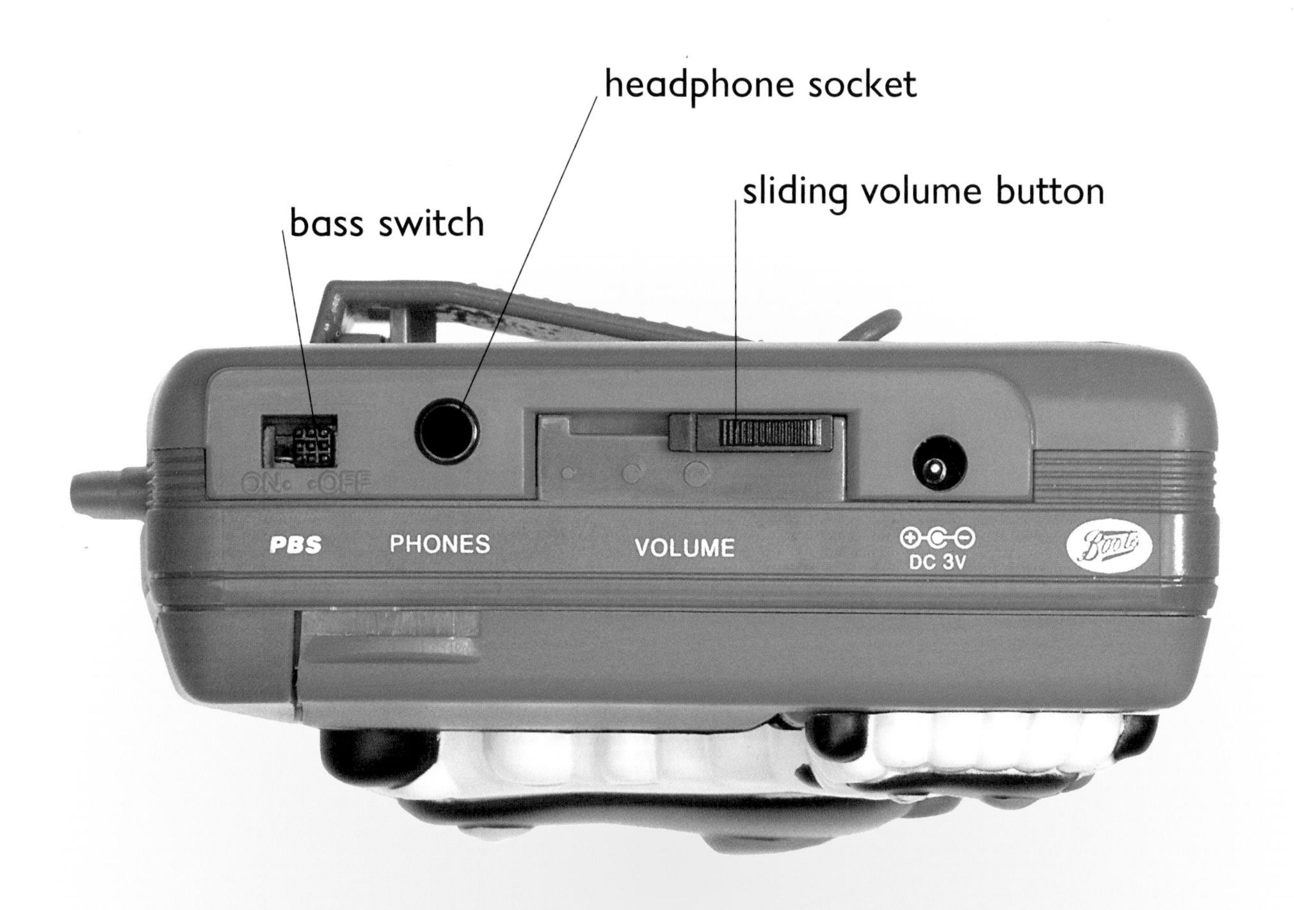

On one side of the cassette player there is a small square switch. This makes **bass** sounds louder. Next to it, there is a round **socket** hole. The plug for the headphones fits into it. You can make the sound loud or **soft** by sliding the **rectangular volume** switch.

THE TAPE

The cassette case is made of hard plastic. It protects the thin plastic ribbon inside. The case is **transparent**. You can easily see if the ribbon gets twisted. The ribbon is coated with a chemical. This holds the sound pattern recorded onto it.

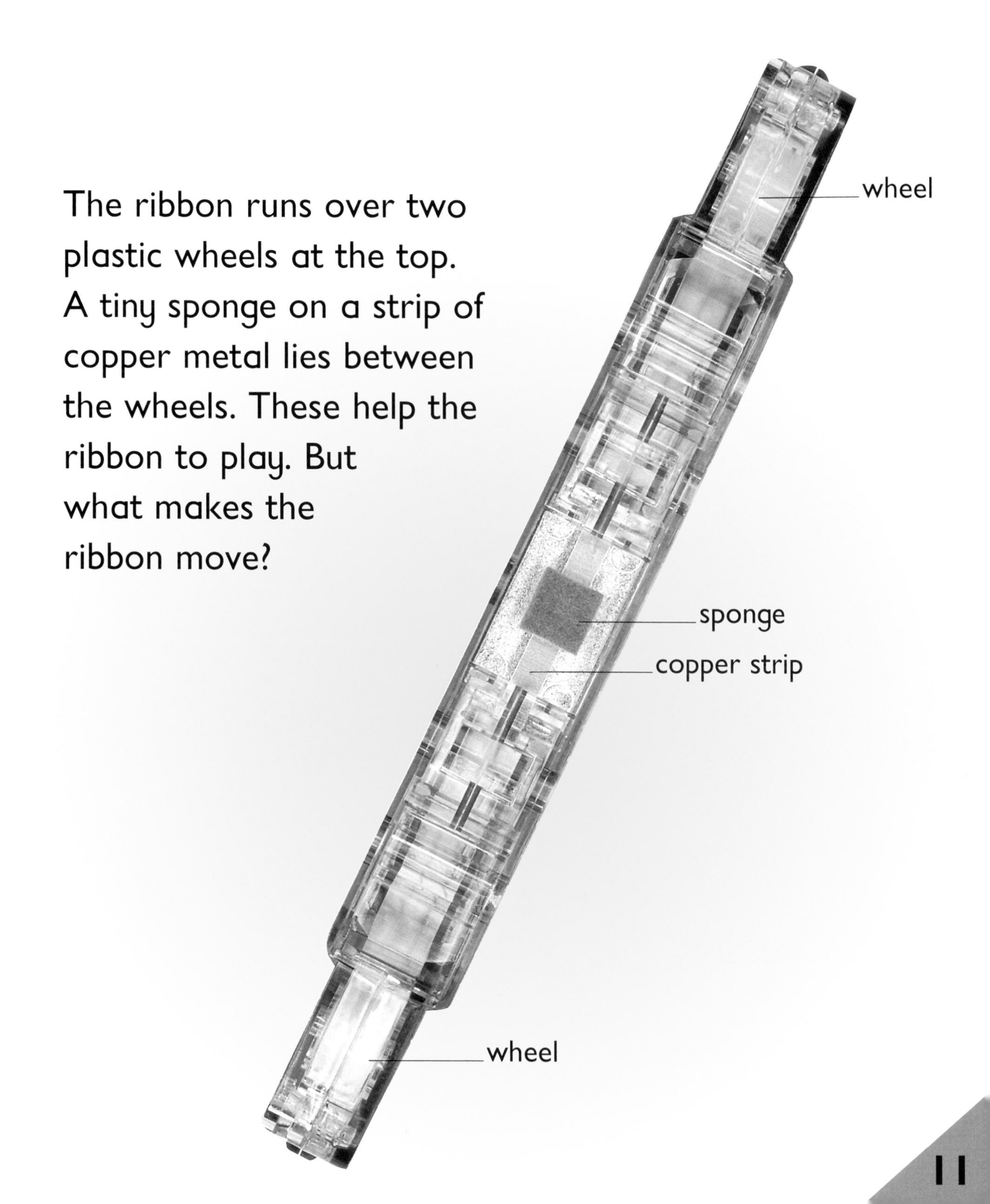

The ribbon runs over two plastic wheels at the top. A tiny sponge on a strip of copper metal lies between the wheels. These help the ribbon to play. But what makes the ribbon move?

MAKING IT MOVE

Power from batteries makes the ribbon move. Inside the battery there is a chemical paste or jelly. The chemicals make electricity when the battery is switched on. The electricity flows through wires to a green **circuit board**. This controls the flow of electric power.

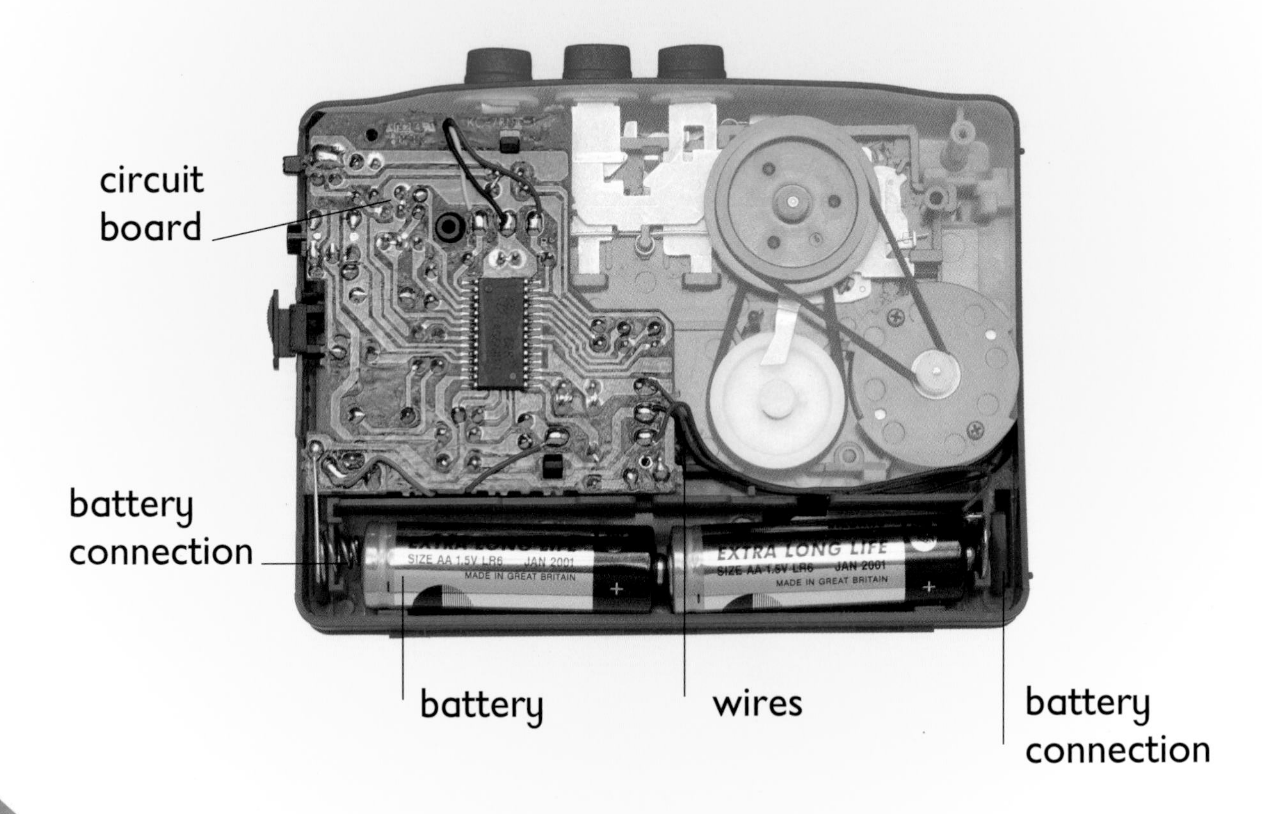

The power turns round the motor. At the back of the motor there are two stretchy rubber bands. These are the drive bands. They drive round two plastic discs, which turn pegs. The pegs are called spindles.

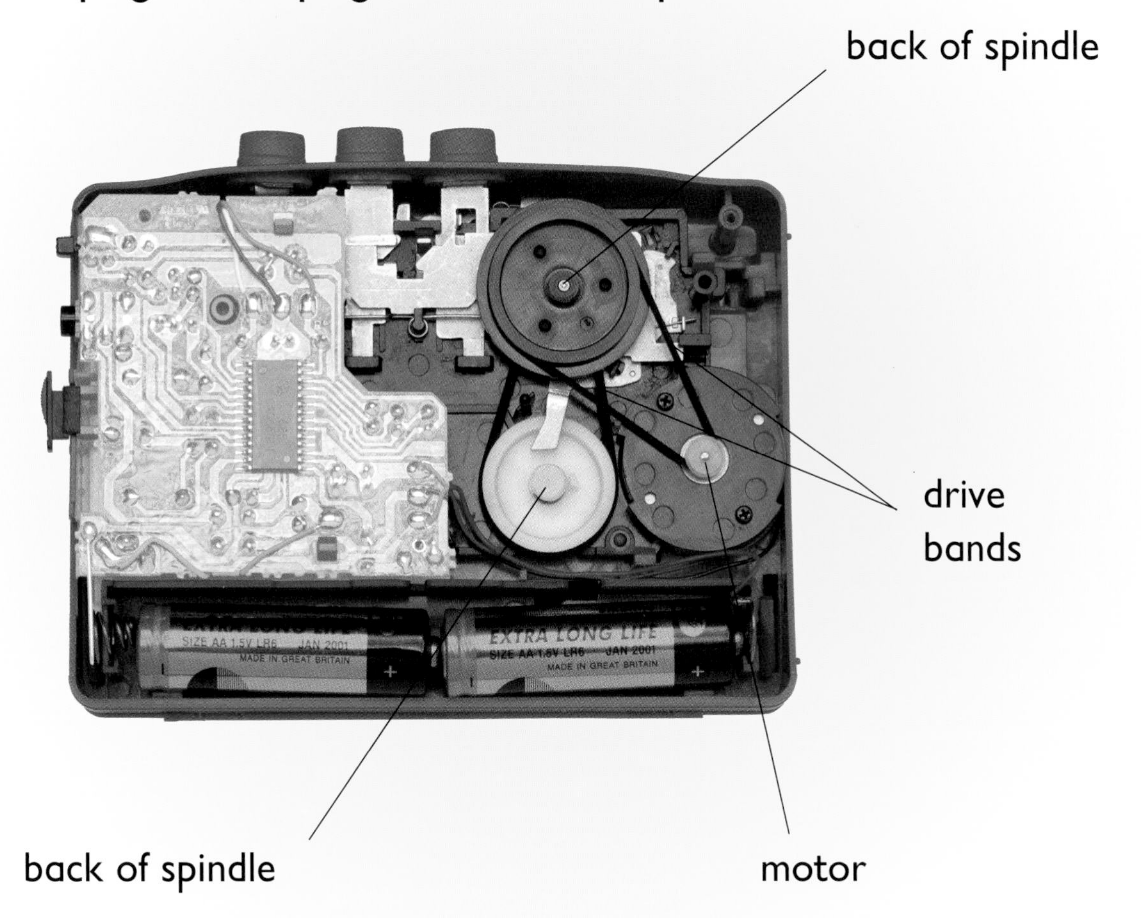

PLAYING THE TAPE

The tape case slots over two spindles in the tape machine. **Grooved** rings cling onto the two black plastic spindles as they turn. This makes the ribbon **reel** across from one side to the other.

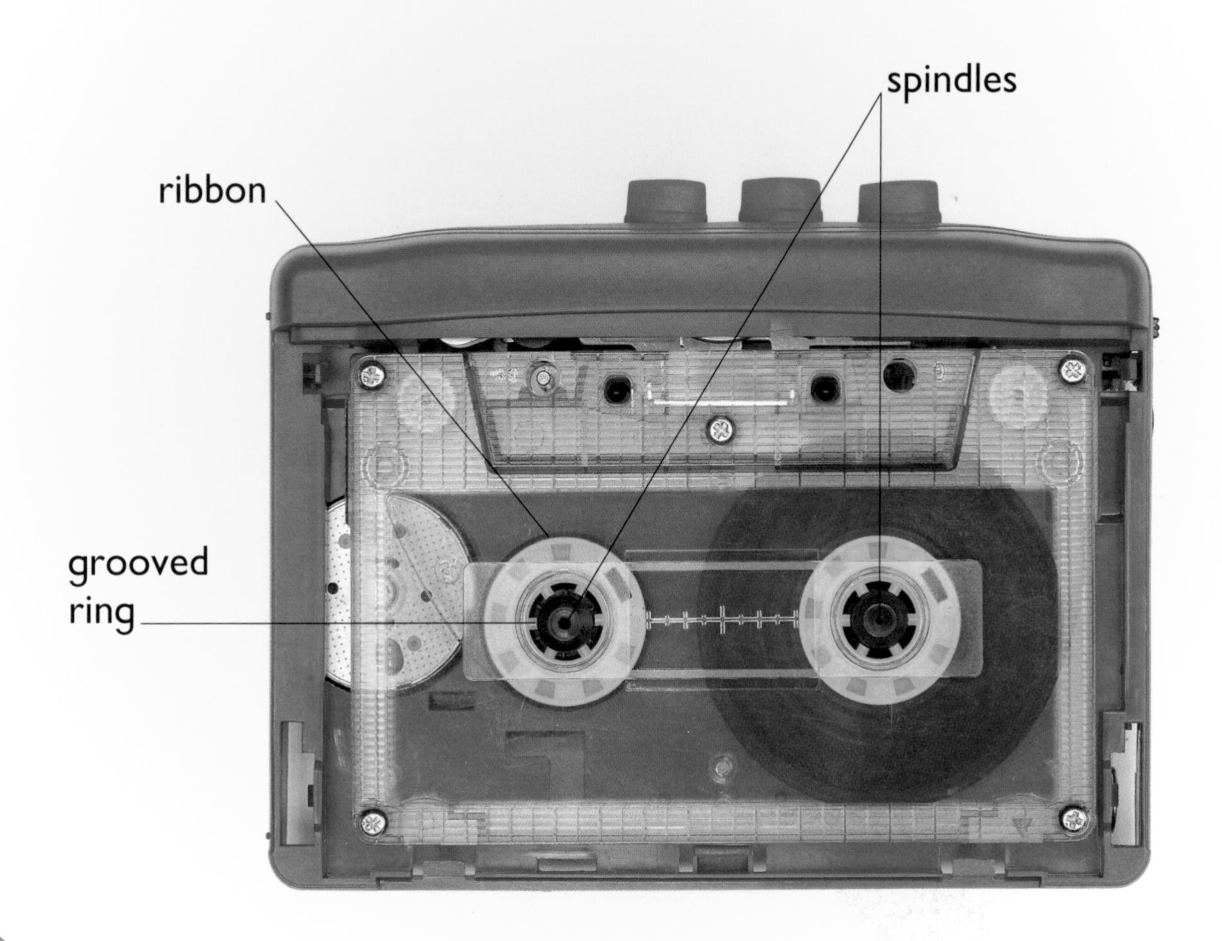

At the top of the tape, a spindle squashes the ribbon against a rubber **pinch wheel**. The spindle moves the ribbon towards a small metal box and the copper strip in the tape case. Now find out how this box helps to make the sound!

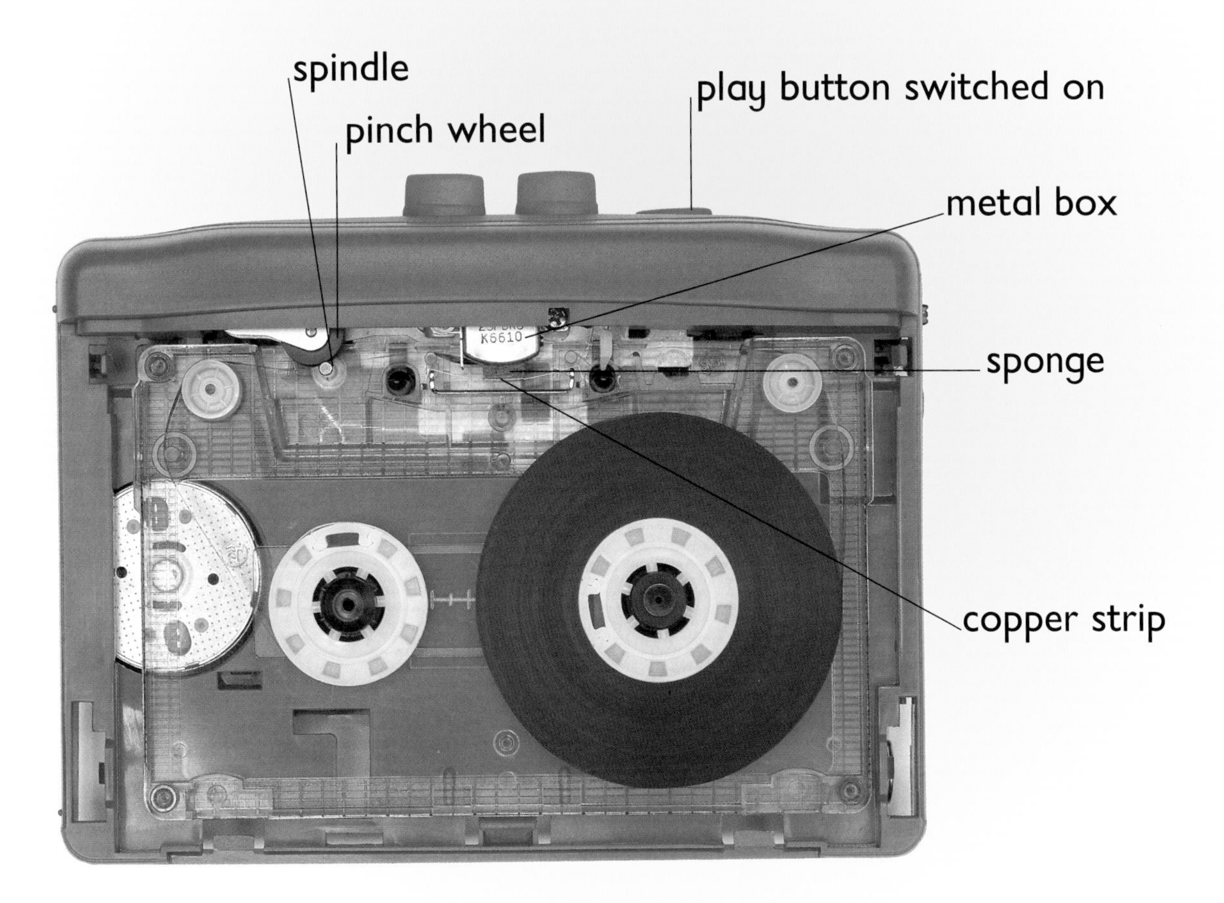

MAKING THE SOUND

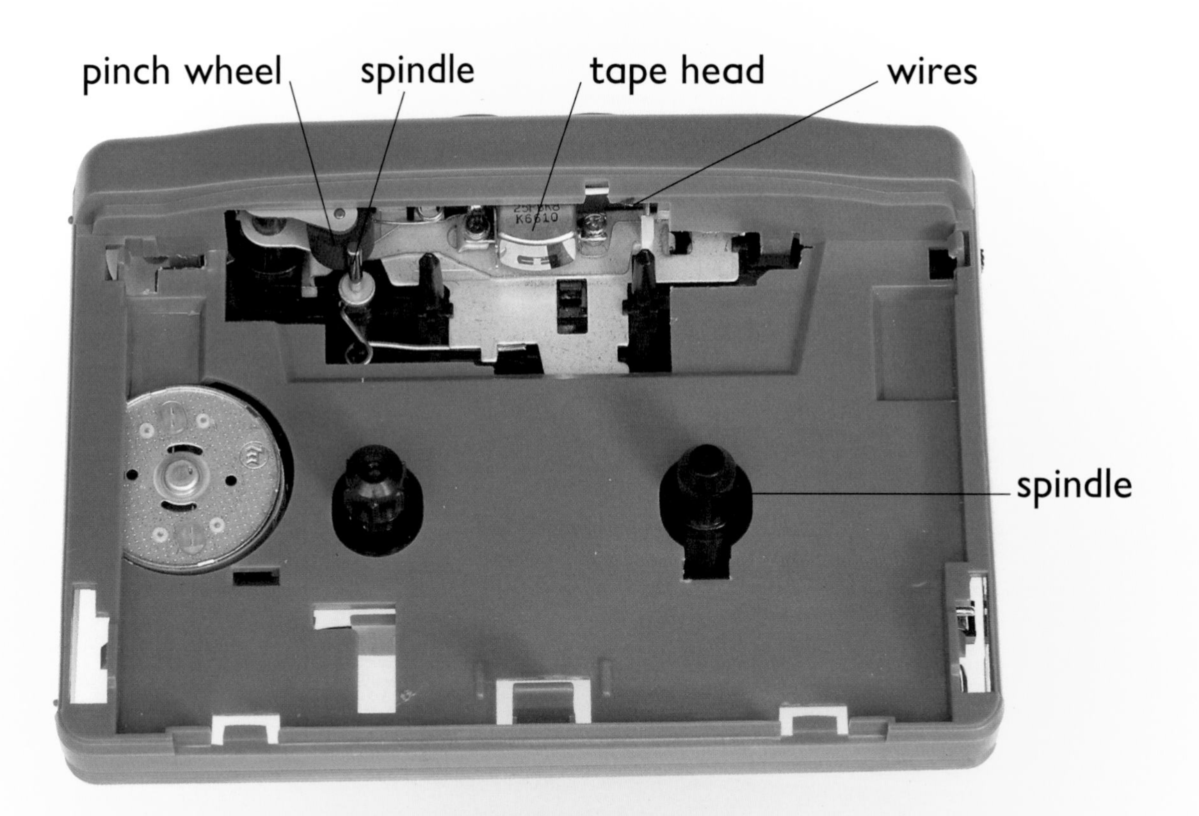

The ribbon squeezes between the copper strip in the tape and the small metal box. The box is called the tape head. Inside there is a tight **coil** of wire. This coil changes the pattern of sound into **electric signals**.

The signals pass through wires to the **circuit board**. Here, the electric signals are amplified. This means they are made really strong. The signals then pass through **flexible** wires to the headphones.

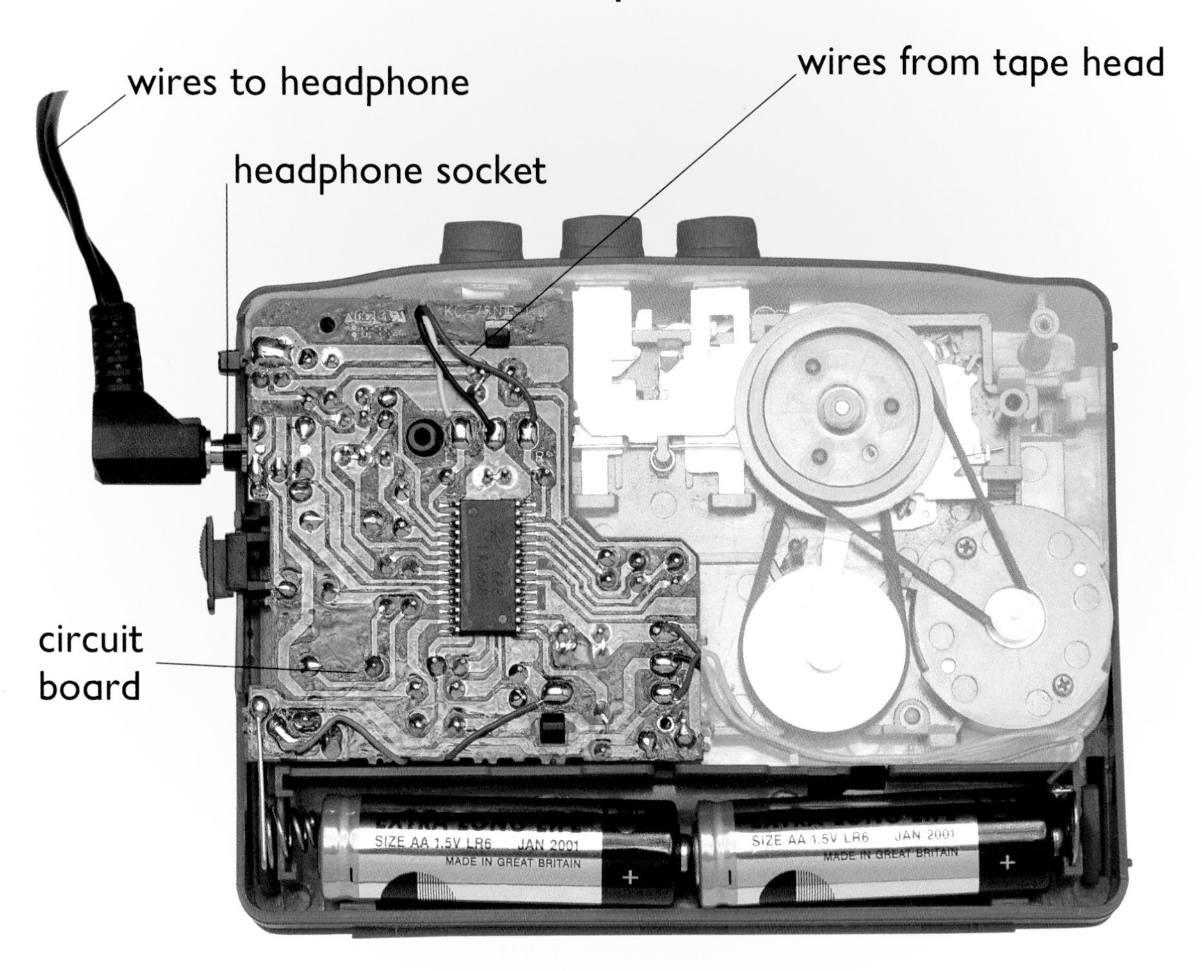

HEADPHONES

Inside the headphones there is a tiny wire **coil**. The strong **electric signals** move the coil. The coil **vibrates** very quickly against a paper-thin **cone**. The cone moves the air. This makes waves of sound. Now you can hear your music through the headphones.

headphone wire

headphone

The headphones are fixed to a band of metal. The metal strip slides to make them bigger or smaller to fit your head. Soft sponges cover the headphones – and your ears! They turn on **hinges** to make them fit well.

USING A PERSONAL STEREO

Here is your personal stereo, all put together. Now open the front of the case and slot in your tape. Next push the play button. Is the sound too **soft**? Then slide the **volume** control to make it louder.

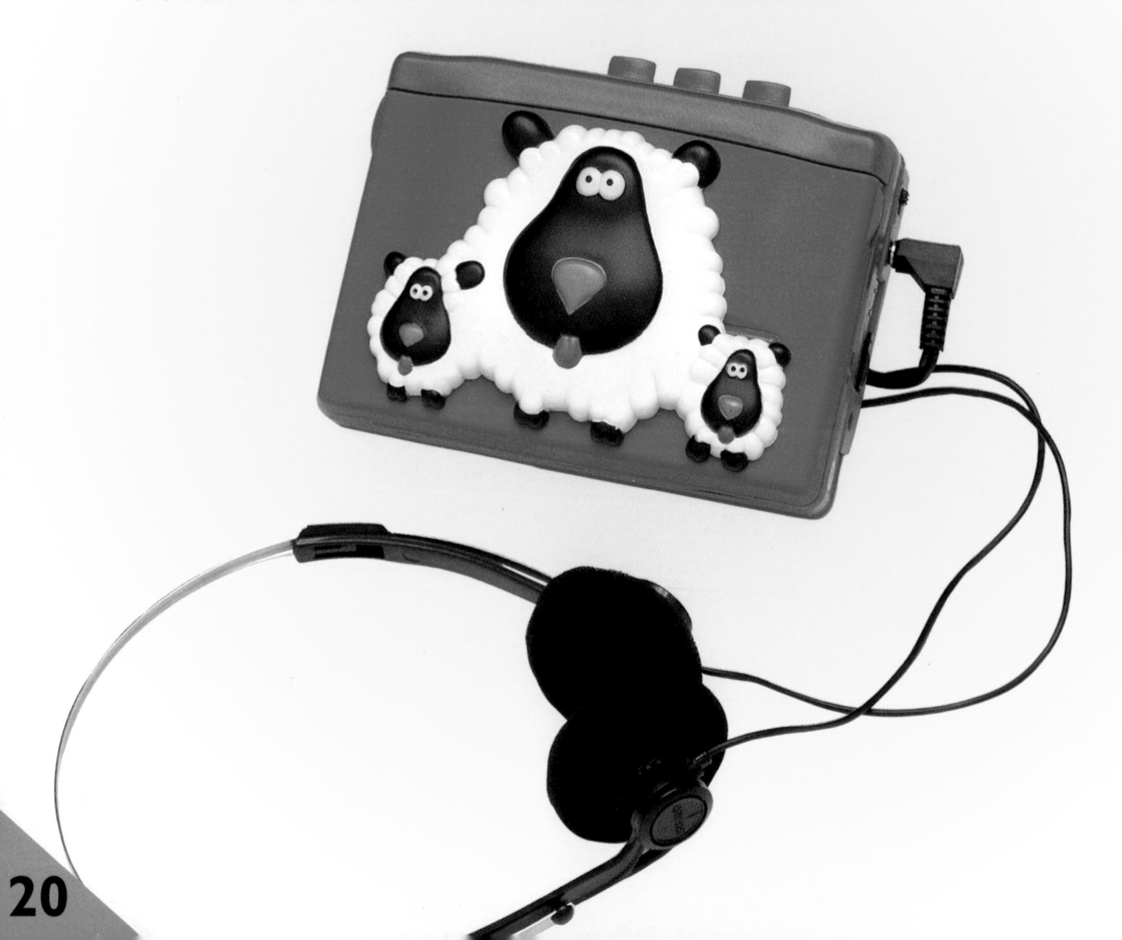

Your personal stereo is very small. It can fit easily on your belt. Or even in your pocket. But it is colourful and fun. You want people to see it. But you don't want anyone else to hear it!

GLOSSARY

bass a low deep sound

circuit board a board that holds tiny switches and wires. These control the flow of signals from one part of a machine to another

coil something that winds round and round, such as a spring

cone a disc that rises to a point in the middle

electric signals messages produced by electricity

flexible bends easily

grooved cut with deep lines

hinges joints which let a door or gate open and close

pinch wheel a wheel that squashes against something

rectangle a shape with four sides

rectangular something that is shaped with four sides

reel to move round and round

socket a hole made for a plug

soft quiet

transparent clear – you can see through it

vibrates moves backwards and forwards very quickly

volume the loudness or softness of sound

Further reading

How Things Work. Steve Parker. Kingfisher Books, 1990

Science and Technology – A Visual Factfinder. Brian Williams. Kingfisher Books, 1993

Science and Technology – Black Holes to Holograms. Oxford University Press, 1993

INDEX